T0365352

From MY NICU WINDOW

I See the Whole World Just for Me.

AMIRIS SABO

ISBN: Softcover 978-1-5245-6017-1
Hardcover 978-1-5245-6016-4
EBook 978-1-5245-6015-7

Print information available on the last page

Rev. date: 11/16/2016

To order additional copies of this book, contact:
Xlibris
1-888-795-4274
www.Xlibris.com
Orders@Xlibris.com

Thank you:

To God; because of who he is. Because he has given me the experience so I can help others with my story and the strength to be able to talk about it.

My husband for supporting me in all my crazy dreams, for being with me in this journey every step of the way and for sharing with me all of the pain and suffering as we went through all this situations.

To my son that has been my inspiration to create this book and this special project. Because he gives me hope every day and when I think I cant continue he is always there to remind me of the greatness of God.

To my illustrator (Iulia Dumitru) that put her heart into this project to draw a beautiful story

And to the friends that in some way have contribute with their advise and time.

Thank you from the bottom of my heart.

Preface

I was inspired to write this book to give comfort to all the families that somehow have been affected by having a child in the NICU (Neonatal intensive-care unit). Just to know that your baby will stay in the Hospital without you it's a lot of stress and pain. To see your child connected to so many things, not knowing if your baby will actually make it home. It is as much my story as it is the story of thousands of families in the world.

My son inspired me to write this book. He was born at 30 weeks gestation from a condition call preeclampsia and spent two months and 3 weeks in the NICU.

Preeclampsia is a condition unique to human pregnancy. It is diagnosed by the elevation of the expectant mother's blood pressure usually after the 20th week of pregnancy. (In my case my blood pressure starts going up from the beginning of the pregnancy). This condition not only affects the unborn baby but also brings new development of decreased blood platelets, trouble with the kidneys or liver, fluid in the lungs, or signs of brain trouble such as seizures and/or visual disturbances for the mother. Some of the symptoms are: headaches, abdominal pain, shortness of breath or burning behind the sternum, nausea and vomiting, confusion, heightened state of anxiety, and/or visual disturbances such as oversensitivity to light, blurred vision, or seeing flashing spots or auras. Preeclampsia and related hypertensive disorders of pregnancy impact 5-8% of all births in the United States.

The subject of preeclampsia is not something that is talked about enough and it is also a reality for many families in the world. This book is not just to give comfort to families with kids in the NICU but also to create awareness about preeclampsia.

The Story

When my husband and I got married we wanted to have children right away. I am pretty sure that lots of couples want the same. We waited for 6 months to get pregnant but at 20 weeks gestation our baby died. It was the worst pain that any parent could feel, his name was Kareem the 1st. Four months later we were expecting again but it only lasted 6 weeks and I had a miscarriage. We waited about 7 months and again I got pregnant, this time we took all of the precautions. We were sure that this time everything will be fine, but at 18 weeks I was admitted to the hospital. After 8 weeks in the hospital with HELLP syndrome (Hemolysis, Elevated Liver enzymes, Low Platelet count), strokes in the placenta my amniotic fluid was going low; my liver and kidneys were failing; My blood pressure was impossible to control and they had to do an emergency C-section and on September 2011 Karimah-Daniela was born. She was extremely premature and sadly past away 4 days later. After we lost her and went through so much I made it my job to give comfort to other families that are going or that went through the same thing. A year after we were expecting our son. I soon as I got pregnant my blood pressure started going up so I was on bed rest for 7 months under blood pressure medication and complete rest. Two times per week I had to get checked until there was nothing else to do and they took me to the hospital at thirty weeks. My blood pressure was sky high and the baby wasn't growing so we waited one week until he had a little more chance of survival and he was out after an emergency C-section. He was 2.8 pounds and 14 inches long (not bad for a preemie). Since he was in the NICU I always notice that he was always very alert and looking every where and I always wonder what all of the babies in the NICU think about. Now in 2016 he is 3 years old and perfect. I wanted to share my story with so many families that are experiencing this, the ones that are feeling proud like me, and the ones that have lost their little ones like I did. I want you to know that you are not alone.

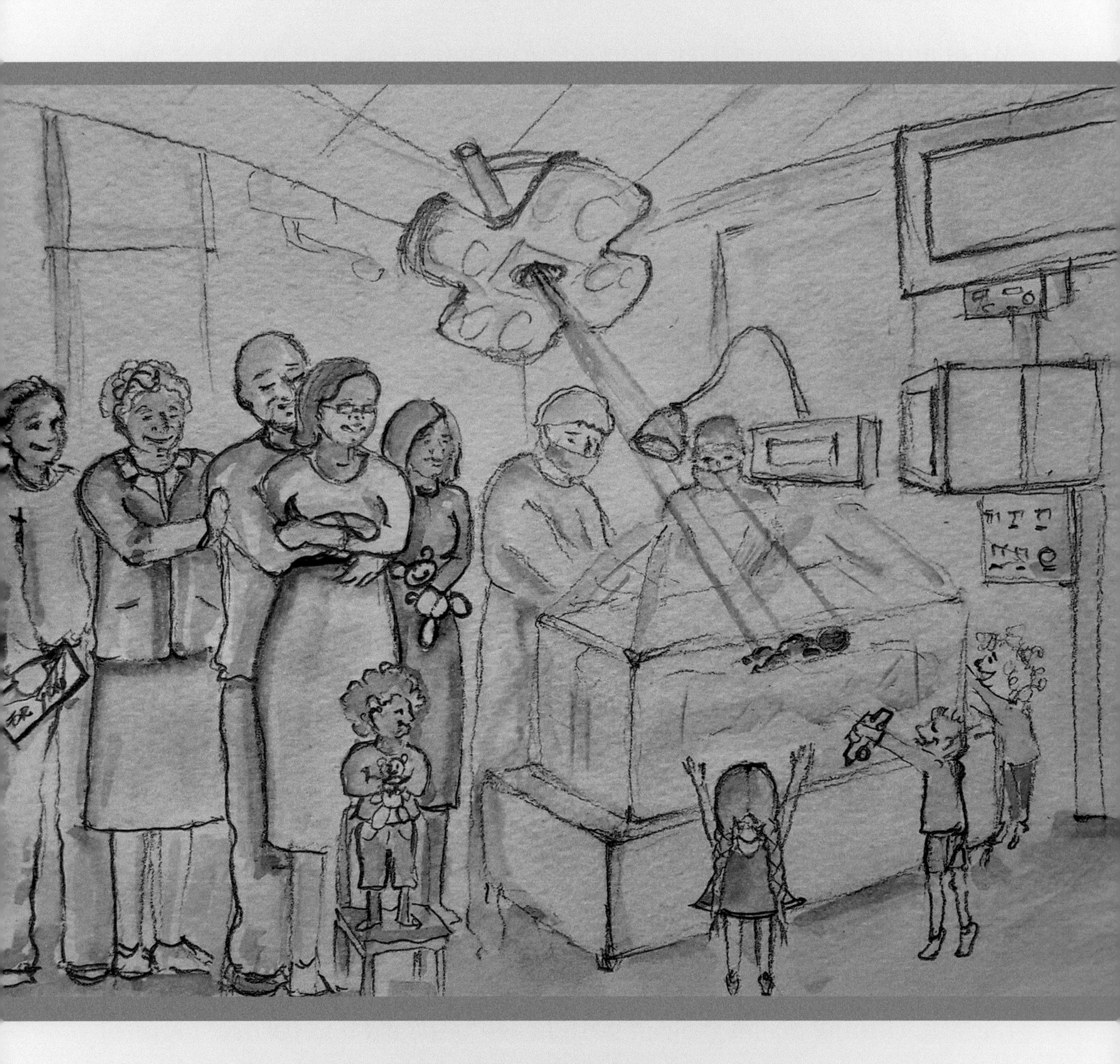

I see my parents, nurses and Doctors and a room full of friends that I could play with.

Veo a Mis padres, Doctores, enfermeras y en mi habitacion muchos niños para jugar

Everything is so big and I am so tiny but full of strength to fight and leave all this behind me.

Todo es tan grande y yo tan pequeño
pero lleno de fuerzas para dejar
todo esto y seguir adelante.

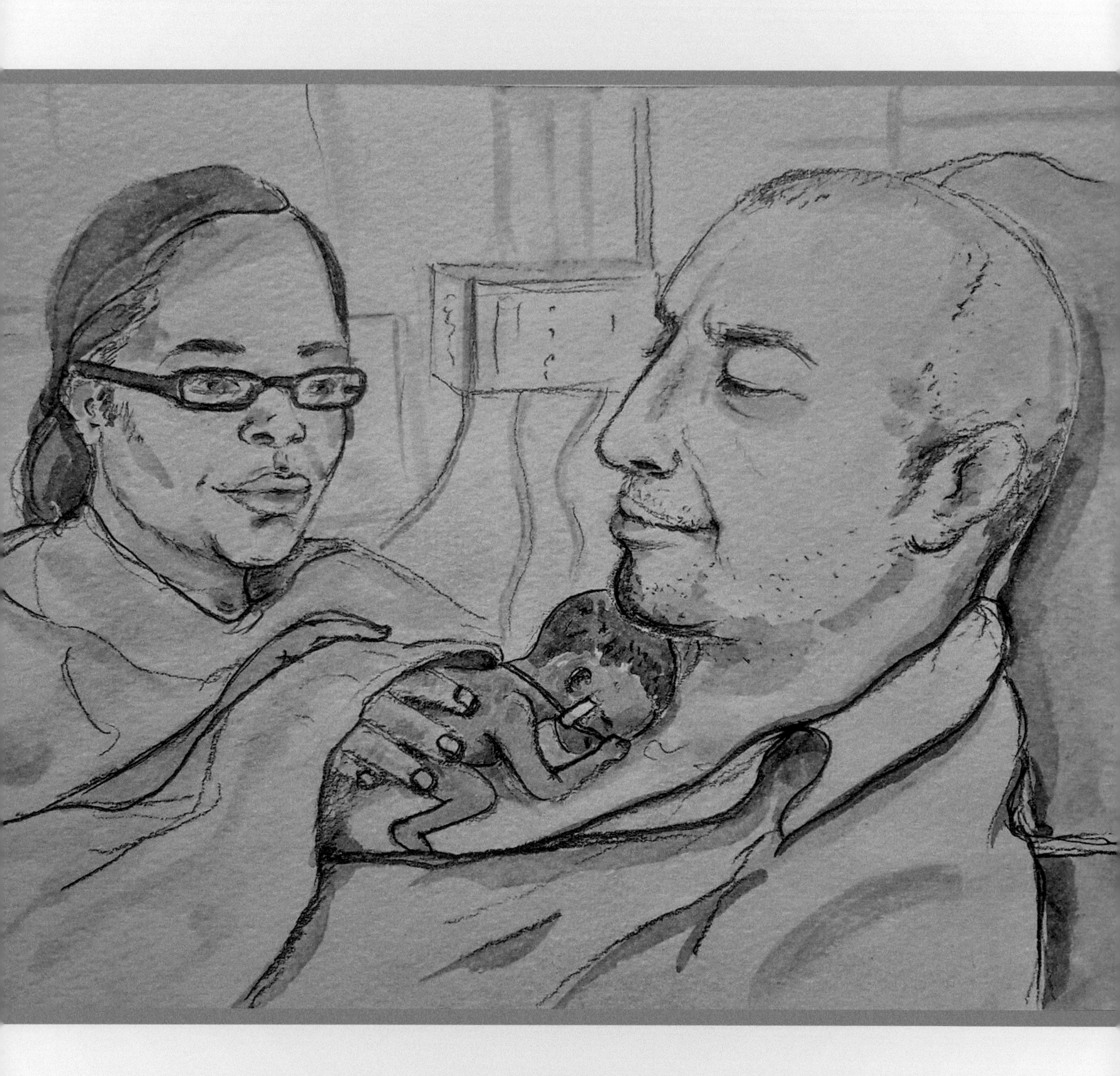

It’s hard to understand, the sounds that my tiny ears can hear, when my mommy cries or my Daddy weeps because of the news they have received.

Los sonidos que mis pequeños oidos escuchan son dificiles de entender, cuando mi mami llora o mi papi esta triste por las noticias que han recivido

I also hear laughs and moments of joy, when my parents hear that I might soon go home.

Tambien escucho momentos felices cuando a mis padres le dicen que puede que pronto me vaya a casa

To go home is a blessing that I am waiting for because some of my friends were not able to go.

Ir a casa es la bendicion que espero ya que algunos amigos no lograron ir.

Buzz

The lights on my bed fascinate me
so much but my parents stay quiet
every time they hear a buzz.

Me facina ver las luces de mi cama pero mis
padres callan cada vez que las escuchan

NICU HOUSE
LOVE
KAREEM
FOR YOU
LIFE

So happy I am in my little NICU house, where I have my friends, my parents, my staff. The nurses take care of all of my needs because I might be tinny but my needs are big.

Estoy tan feliz en mi pequeña casita NICU, tengo a mis amigos, mis padre, y Doctores. Mis enfermeras me dan todo lo que necesito, porque yo sere pequeño mis nececidades son grandes

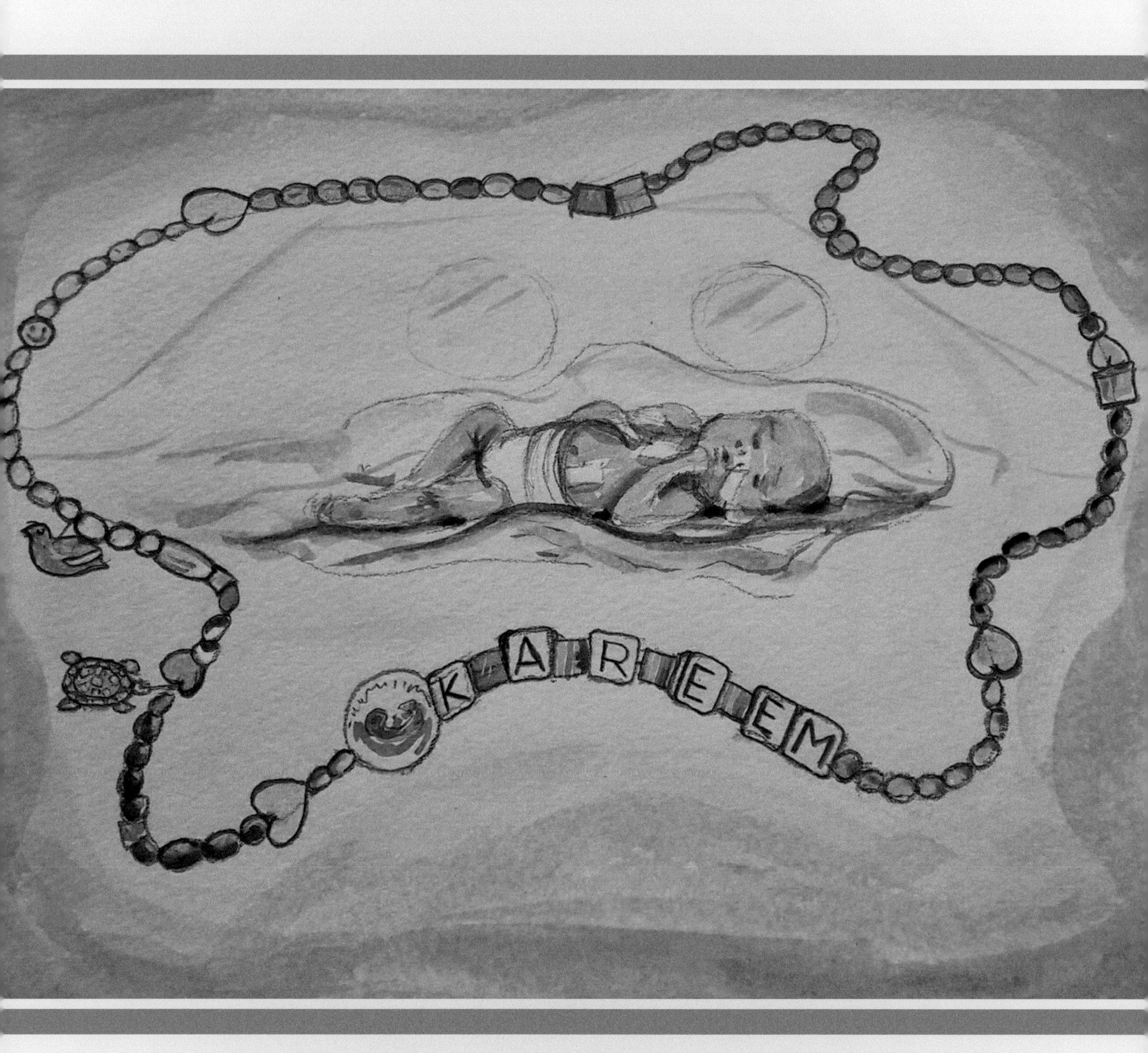
KAREEM

They gave me a necklace; they say is for strength, it shows all of my struggles, victories and fame. It has lots of colors, the stones and a bird, and I love it because it also has my name.

Me Dieron un collar, dicen que me dara fuerzas, muestra mis luchas, victorias y fama. Tiene muchos colores, piedras y un pajaro. Lo amo porque tambien lleva mi nombre

From my NICU window what else do I see? I see a whole world just for me

De mi ventana de el NICU que mas puedo ver? Un mundo entero solo para mi

YOUR STORY

Please share your story with us. If you know a story of survival in the NICU. We will love to hear it.

Email us at HYPERLINK "mailto:Amy58_2000@yahoo.com"Amy58_2000@yahoo.com or on facebook look for our name: "from my NICU window"

Printed in the United States
by Baker & Taylor Publisher Services